CALUMET CITY PUBLIC LIBRARY

3 1613 00405 5813

W9-CZK-564

I n 1917, when Grace Fryer was hired by the U.S. Radium Corporation to work in its Orange, New Jersey, plant, she had no idea that her life was in danger. At the time, Undark, U.S. Radium's trade name for the glow-in-the-dark mixture of radium (chemical symbol: Ra) powder, water, and glue that Fryer and hundreds of other dial painters used to create wristwatch, clock, and airplane instrument faces, was considered completely harmless. At least it was considered harmless by everyone except those who read scientific journals, which reported that exposure to radiation could cause severe health problems.

As early as 1896, scientific journals contained reports of hair loss and burns to the skin caused by radiation exposure. Newspaper reports in 1903 even described the case of Thomas Edison's assistant, Clarence Dally, who lost an arm—and eventually his life—to injuries he suffered while working with X-rays, a high-energy type of radiation. Radiation is energy given off in the form of waves or tiny particles. Moreover, the radiation from radium is much higher in energy than X-rays and is, therefore, more dangerous.

Fryer and her coworkers (almost all of whom were young women) did not know this information, however. Furthermore, to make painting the luminous dials easier, the women routinely used their lips to shape their camel-hair paintbrushes into sharp points. Fryer estimated that she reshaped

Contents

To Marie Curie and all the other women and men, living and dead, who have dedicated their lives to advancing scientific knowledge—we thank you

Published in 2009 by The Rosen Publishing Group, Inc.
29 East 21st Street, New York, NY 10010

Copyright © 2009 by The Rosen Publishing Group, Inc.

First Edition

All rights reserved. No part of this book may be reproduced in any form without permission in writing from the publisher, except by a reviewer.

Library of Congress Cataloging-in-Publication Data

Lew, Kristi.
Radium / Kristi Lew.
    p. cm.—(Understanding the elements of the periodic table)
Includes bibliographical references and index.
ISBN-13: 978-1-4358-5072-9 (library binding)
1. Radium. 2. Periodic law—Tables. 3. Chemical elements. I. Title.
QD181.R1.L49 2009
546'.396—dc22

                                                                          2008022611

Manufactured in the United States of America

On the cover: Radium's square on the periodic table of elements. Inset: A model of the subatomic structure of a radium atom.

J
546
LEW

Pub. 20.00 5-10

Understanding the Elements of the Periodic Table™

RADIUM

Kristi Lew

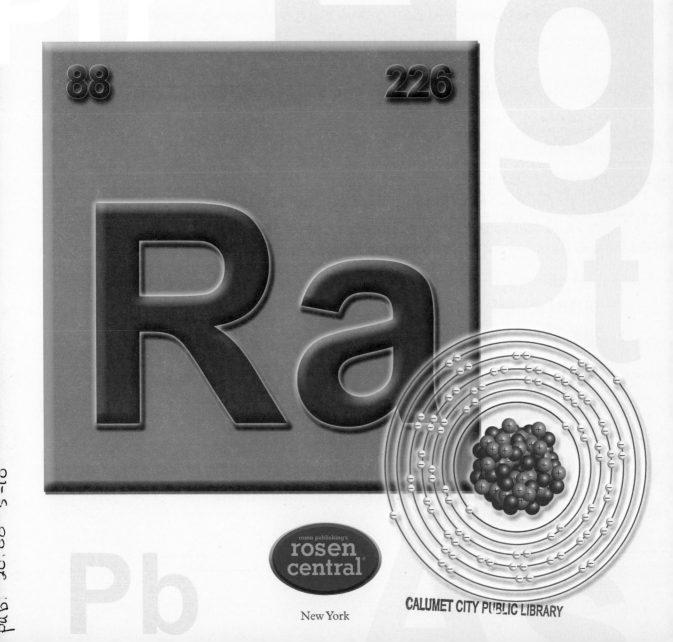

88   **226**

Ra

rosen publishing's
rosen central

New York

CALUMET CITY PUBLIC LIBRARY

Before the health risks of radiation exposure were well understood, young women who worked in clock and watch factories used a mixture containing radium powder to paint glow-in-the-dark numbers on clock and watch faces.

her paintbrush approximately six times for every dial she painted. Most of the women in the plant painted about 250 dials per day. For fun, the women sometimes painted their fingernails and teeth with the mixture, too. This was quite entertaining when they turned off the lights. Later, Fryer would tell the press that the mixture did not taste strange. It had no taste at all, she said. Fryer worked at the U.S. Radium plant for three years. In 1920, she changed jobs and became a bank teller.

Two years later, in 1922, her teeth began to fall out. When she saw a dentist to find out why, he found that the bones in her jaw were also deteriorating. Several years after Fryer's diagnosis, this condition would become known as radium necrosis. Necrosis is the death of living cells or tissue. The cells in Fryer's jawbone were dying because of her exposure to radiation from radium while she worked at the dial-painting plant.

Fryer was not the only U.S. Radium worker suffering from poor health. Between 1922 and 1924, four U.S. Radium employees or former employees died. Workers in the factory began to suspect that the glowing powder they worked with every day had something to do with their coworkers' deaths. They were not the only ones who suspected this powder. Harvard physiology professor Cecil Drinker knew there was a problem at the plant. In June 1924, Drinker discovered that Edward Lehman, U.S. Radium's top chemist, had several serious open sores on his hands. Drinker wrote a report suggesting that the plant should change the way that radium was handled to protect workers like Fryer, Lehman, and many others. Despite Drinker's warning, Lehman and Arthur Roeder, the president of U.S. Radium, insisted that the radium mixture was harmless. Lehman died a year later.

That same year, Fryer decided to take U.S. Radium to court to help pay her medical bills. However, it took her nearly two years to find a lawyer willing to take her case. Finally, Raymond Berry, a New Jersey attorney, agreed to take on U.S. Radium in court. Berry filed a suit against U.S. Radium on May 18, 1927, on behalf of Fryer, Edna Hussman, Katherine Schaub, and sisters Quinta McDonald and Albina Larice. In the press, these women became known as the "Radium Girls."

In June 1928, the Radium Girls settled their case out of court when the company offered to pay each woman $10,000 plus $600 per year for the rest of their lives. This was much less that the $250,000 requested by each woman, but they were sick and dying and the court case was taking a toll on their health. McDonald died the next year. She was thirty-four years old. Fryer and Schaub died in 1933 at the ages of thirty-four and thirty, respectively. Six years later, Hussman died at the age of thirty-seven. The last, and oldest, of the five Radium Girls, Larice, died in 1946. She was fifty-one years old.

Chapter One
Finding Radium

It did not take long for scientists to observe the adverse symptoms of radiation exposure. However, it did take a while for them to recognize that it was the radiation that was causing the health disorders. Marie Sklodowska Curie (1867–1934) and her husband, Pierre (1859–1906), the scientists who discovered radium, for example, had fingers and hands that were burned and hardened because they were so often exposed to radiation. They were also weak and sickly throughout their lifetimes, but they always attributed their illnesses to something other than the radium with which they were working. When Madame Curie died on July 4, 1934, at the age of sixty-six, though, she was suffering from

Marie Curie, who discovered radium, also suffered from health problems caused by exposure to radiation emitted by the element.

leukemia, a type of blood cancer thought to have been caused by her prolonged exposure to radium's radiation. One of Madame Curie's daughters, Irène Joliot-Curie (1897–1956), who was a distinguished scientist in her own right and who continued her mother's investigation of radiation, also died of leukemia, at the age of fifty-eight.

Discovering Radiation

No one knew the effects of radiation exposure on the human body when the German physicist Wilhelm Conrad Röntgen (1845–1923) discovered a new type of ray on November 8, 1895. He was studying how very high-energy electric currents moved though gases under very low pressures. Not knowing what these strange, invisible rays were, Röntgen named them X-rays ("X" being a common mathematical symbol for an unknown). After two months of studying the properties of these new rays, Röntgen realized that they would pass through some materials but not through others. He asked his wife to place her hand in the path of the X-rays, and, in doing so, produced the image of the bones in his wife's hand on a photographic plate, the world's first X-ray image. In 1901, the first Nobel Prize in Physics was awarded to Röntgen for his discovery.

French scientist Antoine Henri Becquerel (1852–1908) was fascinated with Röntgen's discovery. He wondered if X-rays could be produced by light energy as well as by electrical energy. In March 1896, Becquerel decided to study substances that fluoresced (glowed when they absorbed light energy). He wanted to know if these substances would give off X-rays. One of the chemicals that Becquerel tested contained the element uranium (U). Becquerel placed the uranium compound he was studying on top of photographic paper that he had wrapped in heavy black paper. He then put everything in sunlight because bright sunlight caused the uranium compound to fluoresce. If the compound was producing X-rays, Becquerel theorized, then the X-rays should pass through the heavy black paper and fog the photographic paper. His experiment worked.

Yet, before he told anyone about his results, Becquerel wanted to try the experiment again to confirm his results. He set up the experiment, but before he could carry it out, the weather turned cloudy. He placed the uranium compound and photographic paper in a drawer. After several days, he finally got tired of waiting for the sun to come back out and he developed the film anyway. He did not expect to see anything, but to his surprise, the film was fogged just as it was when placed in the sun. From these results, Becquerel determined that the uranium compound was indeed producing some type of invisible rays. Whatever the rays were, however, they were not X-rays. Producing X-rays requires an outside energy source, but these rays did not. The uranium compound was giving off radiation on its own. Becquerel had discovered radioactivity, the spontaneous production of radiation. After finding what he called uranium rays, however, Becquerel decided not to pursue his discovery.

Discovering Radium

Two years later, Marie Curie (who worked in the same laboratory as

The minerals pictured here were stones used by Marie Curie during her experiments. Curie isolated radium from a uranium-containing mineral, called pitchblende. She had to process several tons of pitchblende to get just one gram of radium.

Becquerel) decided to pick up where Becquerel left off. Using equipment invented by her husband, Pierre, Madame Curie was able to measure even the smallest amounts of radioactivity given off by various substances. She discovered that the amount of radioactivity depended on the amount of uranium the substance contained. However, while working with a uranium-containing mineral called pitchblende, she noticed that the mineral was four times as radioactive as it should be for the amount of uranium it contained. From these measurements, Madame Curie deduced that there must be at least one radioactive element other than uranium present in pitchblende.

After a lot of hard work, she was successful in separating two other radioactive elements from pitchblende. These new substances had chemical properties similar to bismuth (Bi) and barium (Ba). Madame Curie had discovered two new elements. She named the first element she isolated polonium (Po) for her native country, Poland. She named the other element radium from the Latin word *radius*, which means "ray."

Extracting radium from pitchblende was not an easy task. To get just one gram of radium, Madame Curie had to process several tons of pitchblende. Even though she discovered radium in December 1898, it would take her

Radium in Nature

Natural deposits of pitchblende can be found in parts of Africa, the Czech Republic, and Slovakia. However, pitchblende is not the only uranium mineral to contain radium. In fact, any rock that contains uranium will also have radium in it. Most of the uranium in the United States is found in Utah and New Mexico. Arizona, Colorado, Utah, and Texas are some of the states that have smaller deposits of uranium-containing minerals.

four more years to isolate enough pure radium to determine its exact atomic weight and its placement in the periodic table. Radium is much more radioactive than uranium, about one million times more. Moreover, the notebooks that the Curies used to record their research are still too radioactive to be safely handled even today.

In 1903, Marie Curie became the first woman to be awarded a Nobel Prize. She shared the 1903 Nobel Prize in Physics with her husband, Pierre, and the French scientist Antoine Henri Becquerel in recognition of their discovery of radioactivity. In 1911, Madame Curie again made history by receiving a second Nobel Prize, this time in chemistry, for her discovery of the elements polonium and radium. It would be thirty-two years before

Radium Springs

Today, scientists know that exposure to radiation can be harmful to the human body. In the late 1920s, though, people were convinced that radiation exposure was not only harmless, it was also healthy. Consequently, when some mineral hot springs, such as the one in Radium Springs, Georgia, were found to contain trace amounts of radium in the water, the springs quickly became popular destinations for tourists hoping to increase their health.

Some resorts in the 1920s advertised radium-containing hot springs. Most hot springs contain only trace amounts of radioactivity that comes from radon, not radium.

another woman would win a Nobel Prize. The next woman to be honored was Madame Curie's daughter, Irène Joliot-Curie, who won the 1935 Nobel Prize in Chemistry for producing new radioactive elements.

Radium on the Periodic Table

Marie Curie determined that radium belonged below the element barium in the second column, or group, of the periodic table. Elements in group 2 (or group IIA in an older naming system) of the periodic table are sometimes called the alkaline earth metals. The periodic table is a tool used by chemists to organize all the known elements. Rows on the periodic table are called periods. On the periodic table, radium is found in period 7.

Elements are substances that cannot be broken down into simpler substances by ordinary chemical or physical means. In other words, an element cannot be changed into another element by heating it, passing electricity through it, exposing it to a strong acid, or any other method that could be carried out in a normal chemical laboratory.

In 1869, a Russian chemist named Dmitry Mendeleyev (also spelled Dmitri Mendeleev; 1834–1907) published one of the first periodic tables. In his version of the periodic table, Mendeleyev arranged the sixty elements known at that time in order of increasing atomic weight. He organized his table so that it listed all the elements with similar properties in the same column. When he came to a spot where no known element had the proper properties necessary to fit into the column at that position, he left a blank. Mendeleyev was convinced that these blanks would eventually be filled in when other elements were discovered with the predicted properties. In 1875, Mendeleyev was proven correct with the discovery of gallium (Ga). Gallium showed properties that fit neatly into one of the blanks he had left in his periodic table. Mendeleyev was proven correct again in 1879 with the discovery of scandium (Sc), and once again when germanium (Ge) was found in 1886.

Henry Moseley, seen here in his laboratory, rearranged the periodic table so that elements are listed by increasing atomic number instead of atomic weight.

The modern periodic table is no longer arranged by atomic weight. In 1913, Henry Moseley (1887–1915), a British physicist, rearranged Mendeleyev's periodic table and ordered the elements by increasing atomic number instead. Radium's atomic number is 88.

Henry Mosley did not just rearrange Mendeleyev's periodic table on a whim, though. He had his reasons, all of which were based on new scientific discoveries about how the atom was put together.

In 1803, English chemist John Dalton (1766–1844) developed an idea called the atomic theory. Dalton's atomic theory stated that all matter is made up of tiny indivisible particles that he called atoms. It also explained that each element is made up of a particular type of atom and all atoms of the same element are identical. An atom is the smallest part of an element that still has the properties of that element. Dalton's theory also included the idea that atoms of two or more elements could combine chemically to form chemical compounds and that during chemical reactions, atoms are rearranged to form different compounds.

Dalton's idea that atoms were the smallest particles of matter lasted for nearly one hundred years. Then, in 1897, English physicist Joseph John Thomson (1856–1940) found the first subatomic particle.

Atomic Structure of Radium

Subatomic particles are particles that are smaller than an atom. There are three main subatomic particles: protons, neutrons, and electrons. The one

J. J. Thomson discovered the first subatomic particle, the electron, in 1897. When chemicals bond to form chemical compounds, they lose, gain, or share electrons.

that Thomson found was the electron. Electrons are negatively charged subatomic particles that travel around the center of an atom on energy levels, or shells. To form a chemical compound, atoms lose, gain, or share electrons. In 1913, Mosley realized that the way elements react with one another had more to do with these subatomic particles than it did with how much the elements weighed. Therefore, Mosley rearranged Mendeleyev's periodic table to reflect the fact that subatomic particles caused the elements to exhibit their particular properties.

Working in the same laboratory as Thomson, another scientist named Ernest Rutherford (1871–1937) began investigating the uranium rays found by Becquerel. Rutherford found that if he covered uranium powder with thin sheets of metal foil, some of the uranium rays would pass though the metal, while others would not. From these observations, Rutherford determined that there were at least two different types of uranium rays. He named the rays that did not pass through the metal foils alpha rays (now called alpha particles). The rays that passed through the foils, he named beta rays (now called beta particles). Rutherford also determined that alpha rays were positively charged, while beta rays were negatively charged.

Rutherford continued to experiment with the alpha rays. In one experiment, he bombarded an extremely thin piece of gold foil with the positively charged particles. The foil was so thin that most of the alpha

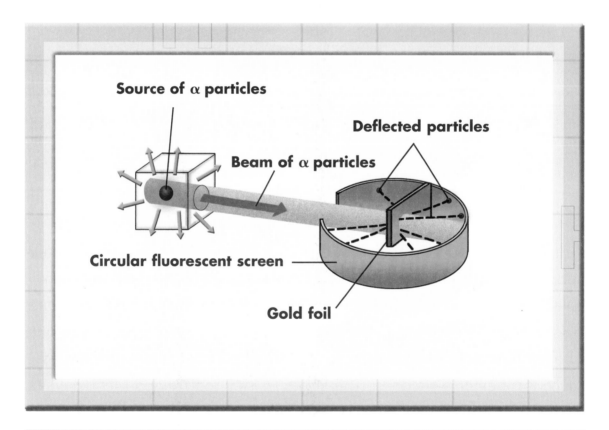

In 1911, Ernest Rutherford discovered the nucleus when he conducted his gold foil experiment. He noticed that when he directed a beam of alpha particles at the foil, instead of passing through it as most particles did, some alpha particles bounced off it.

particles went straight through it. He discovered, however, that some of them bounced back. Rutherford realized that the only way the alpha particles could be reflected back toward the source was if they were hitting something small and heavy inside the gold atoms that made up the foil. Rutherford had discovered the nucleus, or the center, of the atom. The word "nucleus" comes from a Latin word that means kernel, or "little nut."

In 1919, Rutherford found a positively charged particle inside the nucleus of the atom. He named this subatomic particle the proton. Each element has a specific number of protons. Radium, for example, has eighty-eight protons. If an atom has more or less than eighty-eight protons, then it cannot be a radium atom. If an atom has eighty-nine protons, for example, then it is an atom of actinium (Ac). And, if an atom has eighty-seven protons, then it is an atom of francium (Fr).

Atoms are electrically neutral. In other words, the positive charges of the atom's protons must be balanced by the same number of negatively charged electrons. Therefore, all radium atoms also possess eighty-eight electrons. The number of protons in an atom is called its atomic number. Radium's atomic number is 88. In the modern periodic table, all the elements are arranged in order by increasing atomic number.

The third subatomic particle, the neutron, was a little harder to find. In fact, it was not discovered until a student of Rutherford's, James Chadwick (1891–1974), found it in 1932—thirty-five years after the first subatomic particle was discovered. It took longer to find the neutron because this subatomic particle has no charge.

Neutrons may not have a charge, but they do have weight. Neutrons weigh about the same as a proton: 1.67×10^{-27} kilogram. Because the weight of protons and neutrons is so small and difficult to work with when stated in kilograms, chemists generally use a unit called an atomic mass unit (amu) instead. One amu is equal to 1.67×10^{-27} kg. Therefore, one neutron weighs 1 amu, and so does one proton. Neutrons are found in the central, heavy nucleus of the atom along with the atom's protons.

Radium 88 226 Ra Snapshot

Chemical Symbol:	Ra
Classification:	Alkaline earth metal
Properties:	Silvery, soft, radioactive metal that glows in the dark
Discovered By:	Marie and Pierre Curie in 1898
Atomic Number:	88
Atomic Weight:	226 atomic mass units (amu) for the most common atom of radium
Protons:	88
Electrons:	88
Neutrons:	138 (in the most common atom of radium)
State of Matter at 68° Fahrenheit (20° Celsius):	Solid
Melting Point:	1,292°F (700°C)
Boiling Point:	2,084°F (1,140°C)
Commonly Found:	In all uranium-containing rocks

All radium atoms have 88 protons and 88 electrons. Protons are found inside the nucleus. Electrons move around the nucleus on energy levels.

Together, the protons and neutrons make up almost all the weight of an atom. Because electrons weigh so much less than protons and neutrons (9.11 x 10^{-31} kg or 0.0005 amu), their weight makes up only a tiny fraction of the weight of an atom. The atomic weight of a radium atom is 226 amu. Because radium has eighty-eight protons and each proton weighs 1 amu, the protons in radium make up 88 amu of the atom's weight. The rest of a radium atom's weight must come from its neutrons. Therefore, on average, radium atoms have 138 neutrons (because 138 + 88 = 226).

Isotopes

Not all radium atoms have a weight of 226 amu. Some weigh as little as 222 amu, whereas others weigh as much as 228 amu. Because all radium atoms have eighty-eight protons (otherwise they would be different elements), the difference in the weight is caused by a different number of neutrons. Atoms of the same element that have different weights, and therefore, different numbers of neutrons are called isotopes. There are twenty-five known isotopes of radium. The Curies discovered the most common isotope, radium-226, which makes up more than 99 percent of the radium found in nature. The number after the name of the element is the total number of protons and neutrons in an atom of the isotope. This number is called the mass number. Other isotopes include radium-222, radium-223, radium-224, radium-225, and radium-228. The weight listed on the periodic table for an element is an average of the weight of all of that element's isotopes taking into account how frequently they occur. This average is called the atomic weight of the element.

Metal vs. Nonmetal

On the periodic table, radium is the last of the group 2 elements. Like the other elements in this group, radium is a metal. The pure metal is bright silvery-white and its radioactivity causes it to glow in the dark with a faint blue color. When exposed to air, the metal reacts with oxygen and quickly turns black.

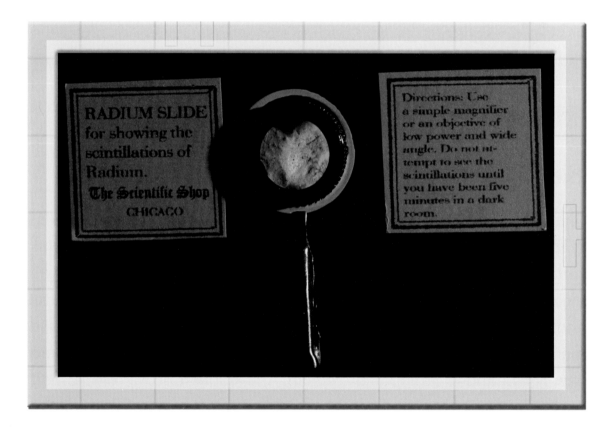

The mixture of radium powder, water, and glue used by Grace Fryer and the other "Radium Girls" to paint clock and watch faces glowed green in the dark, similar to the way this sample glows.

Like most other metals, radium is a solid at room temperature (68° Fahrenheit [20° Celsius]) and is found on the left-hand side of the periodic table. Generally, when metals interact with other elements to form chemical bonds, they lose electrons. Remember that in an atom, the number of electrons is balanced by the number of protons. If electrons are lost, then the atom is left with more protons than electrons and it becomes a positively charged particle. Charged particles that are formed when an atom loses or gains electrons are called ions. If an atom gains electrons, on the other hand, then it has more electrons than protons and forms a negatively charged ion.

A radium atom has eighty-eight electrons in seven energy levels. Eighty-six of a radium atom's electrons are on energy levels one through six. The last two electrons are on the seventh energy level. The electrons on the highest, or outermost, energy level are called valence electrons. In metals such as radium, the valence electrons are lost during chemical reactions. When a radium atom loses its valence electrons, a radium ion with a charge of +2 is formed (because the ion has two more protons than it has electrons). In chemistry, a rule of thumb called the octet rule can be used to predict the number of electrons that an atom will lose, gain, or share during a chemical reaction. According to the octet rule, an atom will lose, gain, or share enough electrons to end up with eight electrons on its highest energy level. The sixth energy level in a radium atom has eight electrons. When the two electrons in the seventh level are lost, the sixth level becomes the highest level and it contains eight electrons.

Nonmetals, on the other hand, usually gain electrons to get eight electrons on their highest energy level. Oxygen, for example, is a nonmetal. Nonmetals are found on the right-hand side of the periodic table. Oxygen's atomic number is eight. Then, an oxygen atom has eight protons and eight electrons. An oxygen atom's eight electrons are found on two energy levels. Two of an oxygen atom's electrons are on the first energy level and six electrons are on its second, and outermost, energy level.

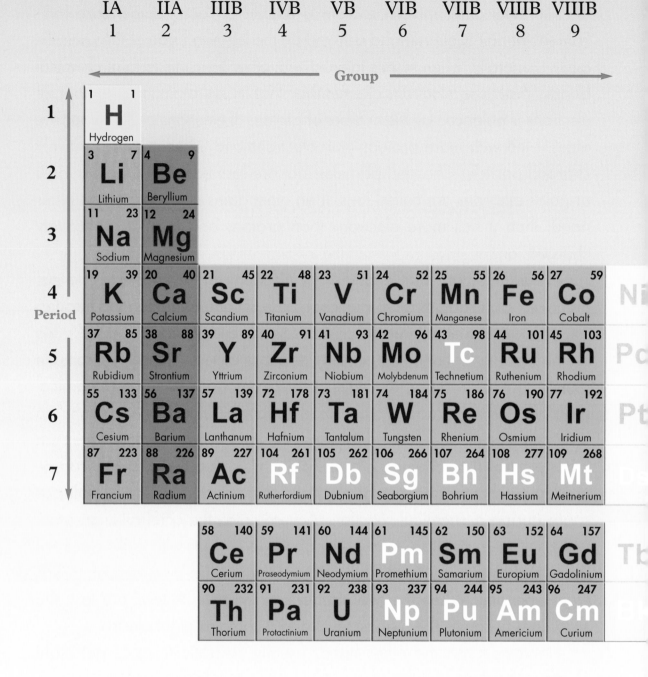

On the modern periodic table, radium is found in period 7 and group 2. Group 2 elements are also called the alkaline earth metals. The alkaline earth metals are the red elements in our chart, and they are generally less reactive than the alkali metals in group 1.

So, an oxygen atom has six valence electrons. For an oxygen atom to conform to the octet rule, it must gain or share two electrons. When an oxygen atom gains two electrons, an oxygen ion with a −2 charge is formed. The attraction between the opposite charges of a positive radium ion and a negative oxygen ion produces a chemical bond between the two ions. This type of chemical bond is called an ionic bond.

The Alkaline Earth Metals

Elements in the same group of the periodic table have similar chemical properties, meaning that they all chemically react in a similar way. Like radium, all group 2 elements have two valence electrons. Therefore, all of the alkaline earth metals form ions with a +2 charge and they are all fairly reactive. Because they react easily, the alkaline earth metals are never found in their pure, uncombined form in nature.

The elements in group 2 react with water, and when they do, they form alkaline solutions. An alkaline solution is one that reacts with an acid. The alkaline earth metals also react easily with oxygen, creating oxide compounds. Early chemists called substances that would not dissolve in water and were unchanged by fire "earths." The oxides of group 2 metals possess many of these properties also. These chemical properties, plus the fact that all group 2 elements are metals, resulted in the group being called the alkaline earth metals.

Chapter Three
Radium and Radiation

Radium differs from all of the other alkaline earth metals in at least one respect. Radium is the only alkaline earth metal that is always radioactive. Unlike other elements, radioactive ones naturally break down and change into different elements. This deterioration of a radioactive element is called radioactive decay.

What Is Radioactivity?

Radioactive decay actually changes the nucleus of the radioactive element. An atom of a radioactive element changes into a different element because the number of protons in its nucleus changes. This happens because the nucleus in a radioactive atom is not stable.

The nuclei (plural of nucleus) in all atoms except hydrogen contain two or more protons. All stable nuclei that contain more than one proton also contain neutrons. Protons all have a positive charge. The positive charge of the protons in a nucleus causes the protons to repel each other. If it were not for the neutrons in the nucleus, then the protons would push each other away and the nucleus would fall apart. The neutrons in the nucleus are like glue that holds the protons together. It takes just the right amount of neutron "glue" to hold the nucleus together. Some nuclei do not have enough neutrons, and others have too many. The nuclei that do not

The nucleus of a radium-226 atom ejects alpha particles to become more stable. The yellow streaks on this specially coated photographic film show alpha particles leaving the nucleus.

have the right amount of neutrons are unstable and they eventually fall apart. That makes these nuclei radioactive.

Some unstable nuclei have too few neutrons. Considered another way, these same nuclei have too many protons for the number of neutrons. These nuclei eventually undergo a change that decreases the number of protons or increases the number of neutrons. One way this happens is for the nucleus to eject protons. However, the nucleus usually does not eject just one. Instead, it ejects a particle that contains two protons and two neutrons. This is an alpha particle. It has a charge of +2 and a mass of 4 amu. It is, in fact, the nucleus of a helium atom, the most stable of all nuclei. Radium-226 decays in this way. When an alpha particle with two protons and two neutrons leaves the radium-226 nucleus, the atomic number goes down by two and the mass number goes down by four. This results in an atom of radon-222.

Some unstable nuclei have too many neutrons (or too few protons, which is the same thing). These nuclei can become more stable by turning a neutron into a proton. They do this by ejecting a particle with a −1 charge and with very little mass, 0.0005 amu. This is a beta particle, which is identical to an electron. The new nucleus has almost the same mass as the old one, but it has a charge of one more than the old one.

3 1613 00405 5813

CALUMET CITY PUBLIC LIBRARY

Radium-228 decays in this way. When a beta particle with a charge of −1 and almost no mass leaves the radium-228 nucleus, the atomic number goes up by one, but the mass number does not change. This results in an atom of actinium-228.

In all radioactive decay, a less stable nucleus turns into a more stable one. Another way of saying this is that the starting nucleus has more energy than the product nucleus. This additional energy is released when one nucleus changes into another. This energy is released in the form of very high-energy radiation known as gamma rays. Gamma rays are not particles. They have no charge and no mass. Instead, they are high-energy waves that travel at the speed of light. In fact, they are like very high-energy light. Gamma rays can easily penetrate the human body. They can only be slowed down and stopped by a thick sheet of lead or slab of concrete.

All radiation is energy moving through space, either as tiny particles or as waves. Some radiation involves only small amounts of energy, such as radio waves and microwaves. Other radiation, such as that in X-rays and from radioactive decay, involves very high energy. High-energy radiation can be dangerous to living things because it has enough energy to break chemical bonds between atoms in molecules. It breaks bonds by knocking electrons out of the molecules, forming ions. For this reason, high-energy radiation is often called ionizing radiation.

All of the radiation from radium, alpha particles, beta particles, and gamma rays is ionizing radiation. Therefore, it is harmful to living things, including people. The alpha particles are the least harmful to people because they are stopped by the outer layer of skin. The outer layer of skin is made of cells that are already dead, so they can't be harmed. Beta particles can make it through the skin and can cause some damage. However, because they are so tiny, they don't do a lot of damage. Gamma rays are the most damaging form of radiation because they

The Argonne National Laboratory's Gammasphere in Argonne, Illinois, is an instrument that helps scientists study unstable atomic nuclei by detecting and analyzing the gamma rays they produce.

Radon Risks

Radon is a colorless, odorless, radioactive gas. It is also a potential health risk. In areas where rocks have a high uranium content, radon gas can seep into people's homes through pipes and drains, cracks in the foundations, or gaps in the homes' insulation. Radon is especially dangerous because it is a gas that can be inhaled. When it is inhaled, it is inside the body, where its radiation is very harmful. Because exposure to radiation given off by radon gas is the second-largest cause of lung cancer in the United States behind cigarette smoking, the Environmental Protection Agency (EPA) recommends that people test their homes for the presence of the gas. Most hardware stores sell do-it-yourself radon testing kits.

carry much more energy than alpha or beta particles and because they can penetrate deeply into the body. As they penetrate, they can break apart hundreds of molecules before they are stopped.

Half-Life

The speed at which a radioactive element decays is expressed by its half-life. The half-life of a radioactive element is the amount of time required for half of the atoms in a radioactive sample to decay into another element. For example, radium-226 has a half-life of around 1,600 years. Therefore, at the end of 1,600 years, only half of the atoms in a sample of radium-226 would still be radium-226. The other half of the atoms would be different elements.

The First Radiation Detector

In 1903, William Crookes (1832–1919) invented the first radiation detector. The device, called a spinthariscope, allowed people to visualize the radioactive decay of radium. It quickly became a popular toy. A tiny amount of radium was placed in the spinthariscope along with a screen that was covered with a chemical compound called zinc sulfide (ZnS). As the radium decayed, alpha particles hit the zinc sulfide, lighting up the screen with what looked like tiny shooting stars (called scintillations).

These spinthariscopes are four of Crookes's originals, which contain radium supplied to him by Marie Curie. Radiation is invisible, but spinthariscope screens are specially coated to show when an alpha particle hits them, making it possible to see the radiation given off by the radium inside.

Each flash of light meant that another radium atom had decayed.

Spinthariscopes are still available today, but the modern devices contain the element thorium (Th) instead of radium. Unlike their predecessors, modern spinthariscopes pose no health risks to the observers of the light show.

Of all the radium isotopes found in nature, radium-226 has the longest half-life. Radium-222 has the shortest, lasting only thirty-eight seconds. The half-lives of radium-223 and radium-225 are measured in days, about 11.5 days and almost 15 days, respectively. Radium-228 has a half-life of 5.76 years. The length of an isotope's half-life gives an indication of the stability of the nucleus of that isotope. Because radium-226 has the longest half-life, it is the most stable of all the radium isotopes.

Decay Products

Radium is found in all rocks that contain uranium. This is because radium is a decay product of that element. Uranium-238 goes through alpha decay to become thorium-234. This is not a very fast process, however, because the half-life of uranium-238 is 4.5 billion years. The decay of uranium-238 to thorium-234 is just the first step in a series of decays that produce other nuclei. After several of these steps,

Uranium-238 Decay Chain

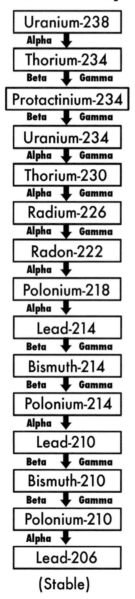

Uranium-238
Alpha ↓
Thorium-234
Beta ↓ Gamma
Protactinium-234
Beta ↓ Gamma
Uranium-234
Alpha ↓ Gamma
Thorium-230
Alpha ↓ Gamma
Radium-226
Alpha ↓ Gamma
Radon-222
Alpha ↓
Polonium-218
Alpha ↓
Lead-214
Beta ↓ Gamma
Bismuth-214
Beta ↓ Gamma
Polonium-214
Alpha ↓
Lead-210
Beta ↓ Gamma
Bismuth-210
Beta ↓ Gamma
Polonium-210
Alpha ↓
Lead-206
(Stable)

Uranium-238 goes through a series of decays to become lead-206, which is a stable isotope. The process begins when uranium-238 ejects an alpha particle from its nucleus to become thorium-234.

radium-226 is produced. Radium-226 emits an alpha particle and becomes radon-222.

The decay sequence does not stop at radon-222. Radon-222 is also radioactive. It changes into polonium-218 by losing an alpha particle. This decay process continues until all the atoms of uranium-238 are changed into a stable isotope of lead. Because the isotope of lead is stable, it is not radioactive and the decay sequence stops. Because uranium-238 starts this sequence, it is called the parent isotope. Thorium, radium, radon, and all of the other decay products are called daughter isotopes.

Chapter Four
The Reality of Radium

Strangely, even though radiation from radioactive elements such as radium and radon can cause cancer, they can also be used to treat the disease. Marie and Pierre Curie were some of the first scientists to visualize how the newly discovered atomic property of radioactivity might revolutionize health care. Madame Curie was convinced that radiation was the key to new scientific discoveries and progress in cancer treatment. She spent many of the last years of her life establishing a research laboratory called the Radium Institute that was dedicated to the study of radiation. Nevertheless, that does not mean that people always used radium and other radioactive elements wisely or safely.

The Radium Craze

Many people in the 1920s believed that radiation exposure was harmless and possibly even healthy. In addition to soaking in radium-laced natural spring waters (which probably did not contain enough radioactivity to actually harm anyone), people added tiny amounts of radium to tea, face creams, lipstick, toothpaste, and bath salts. Many of these products promised to give people a "healthy glow." Cloth bags called "cosmos bags" (named for Henry Cosmos, the man who manufactured them) were also hung around people's necks to cure their arthritis. And people wore glow-in-the-dark costumes painted with radium paint to delight their friends and family.

When some natural hot springs around the world started to become popular tourist destinations because of their radium content, some companies even started bottling the spring water and marketing it under the name of "Radon Water." But because the half-life of radon is only 3.82 days, the chances that there was any radiation left by the time people drank the water is slim. According to the company Radium Ore Revigator, this was a problem. Oh, what were people to do? Never fear, Radium Ore Revigator had a solution—the Revigator. The Revigator was a radium-lined flask that could be filled with water. By using the Revigator, people could make fresh, radioactive water at home!

For anyone who did not want to go to the trouble of making their own radioactive water, the Bailey Radium Laboratories of East Orange, New Jersey, had another solution—a premixed drink called Radithor. Radium water was also called "liquid sunshine." The manufacturer of Radithor assured people that the drink could cure stomach cancer and mental illness. The radium craze lasted for more than a decade and made the people selling these products millions of dollars.

Today, people are much more careful about exposing themselves

Believe it or not, at one time people thought radiation was healthy. The Revigator was a radium-lined flask that enabled people to make their own radioactive water at home.

THE
REVIGATOR
WATER JAR
For Every Home

to radiation. Scientists have proved over the years that this is just not a good idea. Exposure to ionizing radiation breaks chemical bonds in the body. This rearrangement of atoms changes the structure of molecules, including genetic material such as DNA, in the body's cells. Occasionally, a cell can repair itself after it is exposed to radiation. If this happens, then there is no long-term effect on the body. However, cells cannot always repair the damage done by ionizing radiation. In these cases, the cell may be mutated, or changed. Mutated cells generally have several different fates. They can either be detected by the body's immune system and killed, or, if enough damage is sustained, the cell may die on its own. It is also possible that the cell may live on in its mutated state. If the ionizing radiation damages a part of the DNA that tells the cell when to grow and when to die, the cell may not die when it should. In fact, it might grow out of control. Cells that grow out of control cause cancer.

Death of the Radium Craze

One of the people who believed drinking radium-laced water was healthy was a popular American millionaire and amateur golf champion named Eben Byers (1880–1932). Byers drank one to two bottles of Radithor every day for several years. Before he died, parts of his mouth and jaw had to be surgically removed. In fact, Byers's death was reported in a *New York Times* article with the headline: "The Radium Water Worked Fine Until His Jaw Came Off." The radiation poisoning death of Eben Byers prompted the U.S. Food and Drug Administration (FDA) to investigate the safety of radioactive health products. These investigations eventually showed the folly of taking radiation to promote health and put the people manufacturing these products out of business.

Cancer Treatment

It may seem strange that something that can cause cancer could also be used to treat it, but radiation can do both. During cancer treatment, parts of the body that contain cancerous cells are subjected to high doses of radiation. Abnormal cells (like mutated ones that turn cancerous) react especially poorly to radiation. This can kill or damage the abnormal cells. As a result, radiation kills some cancer cells and slows the growth of others. This prevents the cancer from spreading to other parts of the body. Radiation therapy, which is also called radiotherapy, affects normal cells, too. In spite of this, normal cells are better at repairing themselves after radiation treatment than abnormal ones. During radiotherapy, doctors use the lowest amount of radiation possible. This still kills cancer cells, but it affects fewer normal cells. Another strategy that doctors use to kill cancer cells but spare normal ones is spreading radiation treatments out over many months. This schedule allows normal cells to repair themselves between treatments while cancer cells die.

Some types of cancers are treated with "seeds" (small BB-like pellets) or needles that are inserted near the cancerous tissue. This keeps other parts of the body from being affected by the radiation. Some of these seeds contain isotopes of radium. Radium was the first radioactive substance to ever be used to treat cancer. Today, radium chloride is still occasionally used to treat cancer that has spread to the bones from other parts of the body. Nonetheless, radium is highly radioactive and expensive. Consequently, it is used only rarely as the radiation source for radiotherapy or for medical research. Most of the radioactive material used for these purposes today is human-made in nuclear reactors or particle accelerators (also called cyclotrons). In these places, radioactive material can be made safely and more cost effectively.

To be sure, radiation is a dangerous property of some elements. However, handled with care, radiation can also be a lifesaver. Doctors

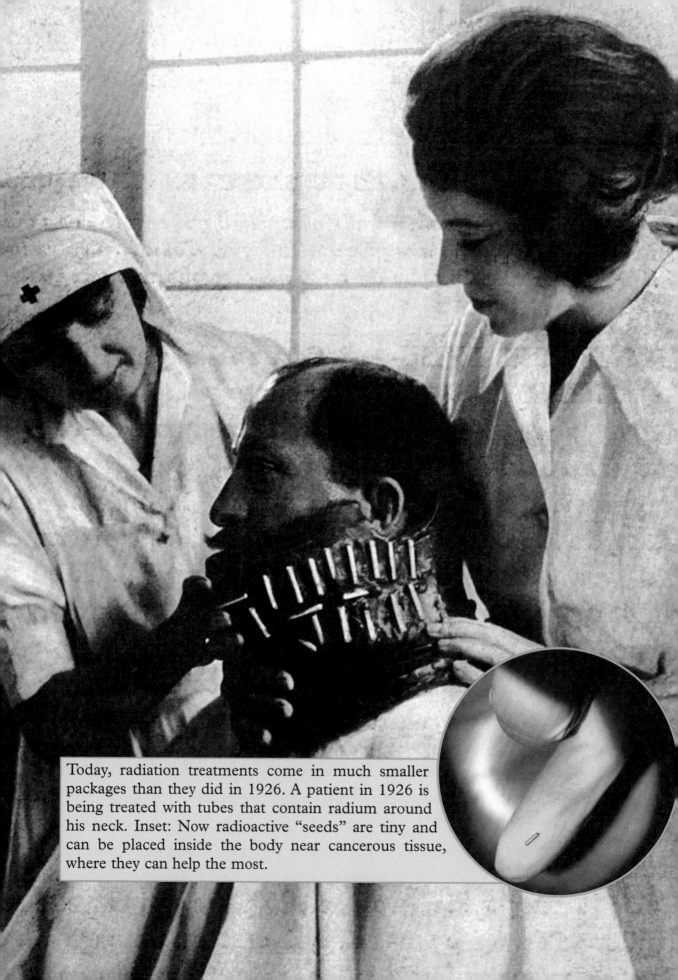

Today, radiation treatments come in much smaller packages than they did in 1926. A patient in 1926 is being treated with tubes that contain radium around his neck. Inset: Now radioactive "seeds" are tiny and can be placed inside the body near cancerous tissue, where they can help the most.

and scientists have learned to use radiation safely and for the benefit of everyone. Radioactivity today is extremely useful in the treatment of cancer and in the production of electricity. Thanks to Marie and Pierre Curie, Henry Becquerel, and many other scientists along the way, doctors and scientists now know a lot more about radiation and how to harness its powerful energy than they ever have before. And they are still learning.

The Periodic Table of Elements

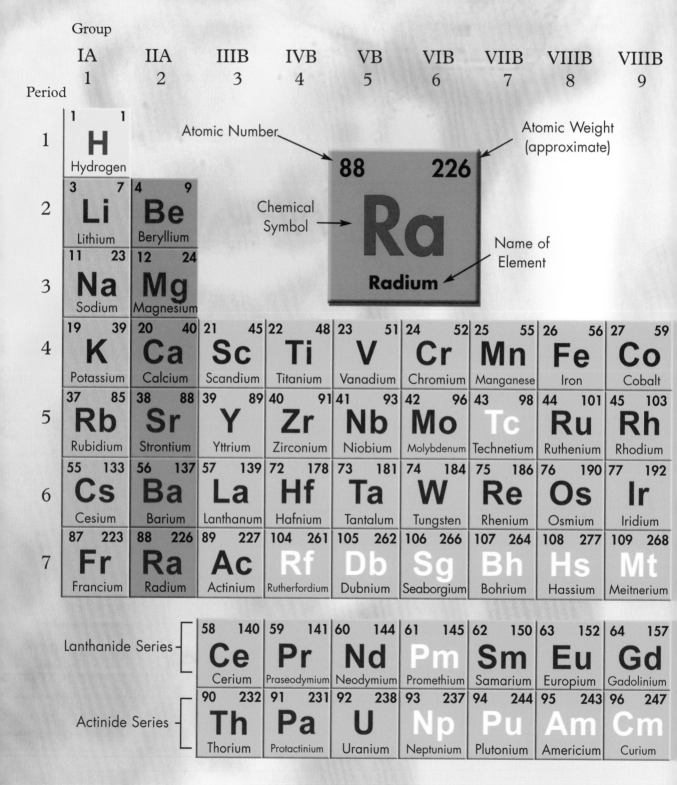

Group

IA	IIA	IIIB	IVB	VB	VIB	VIIB	VIIIB	VIIIB
1	2	3	4	5	6	7	8	9

Period

1 — 1 1 H Hydrogen

Atomic Number

Atomic Weight (approximate)

88 226

Chemical Symbol → **Ra**

Name of Element

Radium

2 — 3 7 Li Lithium | 4 9 Be Beryllium

3 — 11 23 Na Sodium | 12 24 Mg Magnesium

4 — 19 39 K Potassium | 20 40 Ca Calcium | 21 45 Sc Scandium | 22 48 Ti Titanium | 23 51 V Vanadium | 24 52 Cr Chromium | 25 55 Mn Manganese | 26 56 Fe Iron | 27 59 Co Cobalt

5 — 37 85 Rb Rubidium | 38 88 Sr Strontium | 39 89 Y Yttrium | 40 91 Zr Zirconium | 41 93 Nb Niobium | 42 96 Mo Molybdenum | 43 98 Tc Technetium | 44 101 Ru Ruthenium | 45 103 Rh Rhodium

6 — 55 133 Cs Cesium | 56 137 Ba Barium | 57 139 La Lanthanum | 72 178 Hf Hafnium | 73 181 Ta Tantalum | 74 184 W Tungsten | 75 186 Re Rhenium | 76 190 Os Osmium | 77 192 Ir Iridium

7 — 87 223 Fr Francium | 88 226 Ra Radium | 89 227 Ac Actinium | 104 261 Rf Rutherfordium | 105 262 Db Dubnium | 106 266 Sg Seaborgium | 107 264 Bh Bohrium | 108 277 Hs Hassium | 109 268 Mt Meitnerium

Lanthanide Series — 58 140 Ce Cerium | 59 141 Pr Praseodymium | 60 144 Nd Neodymium | 61 145 Pm Promethium | 62 150 Sm Samarium | 63 152 Eu Europium | 64 157 Gd Gadolinium

Actinide Series — 90 232 Th Thorium | 91 231 Pa Protactinium | 92 238 U Uranium | 93 237 Np Neptunium | 94 244 Pu Plutonium | 95 243 Am Americium | 96 247 Cm Curium

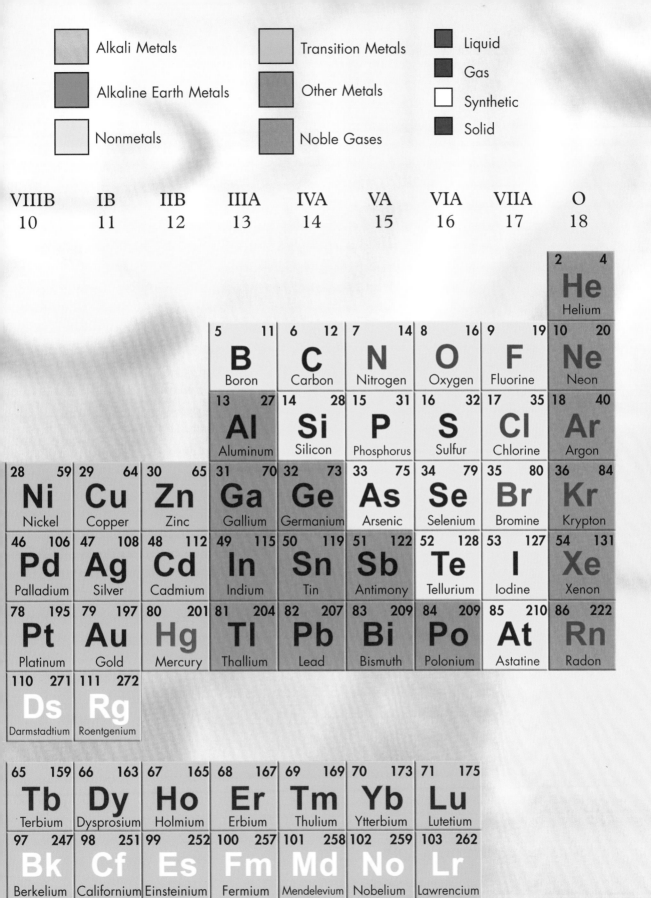

Glossary

alpha particle A particle emitted by some radioactive atoms; it is the same as the nucleus of a helium atom.

atom The smallest part of an element that still has the properties of that element.

atomic number The number of protons in the nucleus of an atom.

atomic weight An average mass (weight) of an atom of an element as found on Earth.

beta particle A particle emitted by some radioactive atoms; it is the same as an electron.

compound A substance that contains two or more elements chemically bonded together.

electron A negatively charged subatomic particle that travels around the nucleus of an atom in energy levels, or shells.

element A substance that cannot be separated into simpler substances by ordinary chemical or physical means.

half-life Amount of time required for half of the atoms in a radioactive sample to decay into another element.

ion A charged particle formed when an atom or molecule loses or gains electrons.

ionic bond The attraction between two oppositely charged ions.

ionizing radiation High-energy radiation that can remove electrons from an atom or a molecule, turning it into an ion.

isotopes Forms of the same element that have a different number of neutrons and a different atomic mass.

mass (or weight) The actual mass (weight) of a particular atom (in grams or atomic mass units).

mass number The total number of protons and neutrons in the nucleus of an atom.

matter Anything that has mass and takes up space.

neutron An uncharged particle found inside the nucleus of an atom.

nucleus The center of an atom that has a positive charge and contains almost all of the mass of the atom.

proton A positively charged particle found inside the nucleus of an atom.

radiation Energy given off in the form of waves or tiny particles.

radioactivity The spontaneous production of radiation from the decay of the nucleus of an atom.

valence electrons The electrons on the highest, or outermost, energy level of an atom.

Argonne National Laboratory
9700 South Cass Avenue
Argonne, IL 60439
(630) 252-2000
Web site: http://www.anl.gov
As one of the U.S. Department of Energy's largest research centers, Argonne National Laboratory provides several K–12 educational programs, including Ask a Scientist, Question of the Week, and information about Rube Goldberg Machine contests.

Atomic Energy of Canada Limited (AECL)
2251 Speakman Drive
Mississauga, ON L5K 1B2
Canada
(866) 886-2325
Web site: http://www.aecl.ca/Resources/Teachers.htm
The AECL sponsors many educational programs throughout Canada, including a list of speakers and a six-week summer program for high school students that allows them to spend time in leading Canadian laboratories.

National Institute of Environmental Health Sciences (NIEHS)
P.O. Box 12233
Research Triangle Park, NC 27709
(919) 541-0395
Web site: http://niehs.nih.gov

The NIEHS is an organization that studies the effects of the environment on human health. The Web pages explain environmental health topics, including the positive aspects of uranium and radiation (http://kids.niehs.nih.gov/uranium.htm).

U.S. Environmental Protection Agency (EPA)
Ariel Rios Building
1200 Pennsylvania Avenue NW
Washington, DC 20460
(202) 272-0167
Web site: http://www.epa.gov
The EPA provides fact sheets and an interactive Web site called RadTown USA that discusses what radiation is, where it comes from, and how it affects people.

Web Sites

Due to the changing nature of Internet links, Rosen Publishing has developed an online list of Web sites related to the subject of this book. This site is updated regularly. Please use this link to access the list:

http://www.rosenlinks.com/uept/radi

For Further Reading

Barber, Ian. *Sorting the Elements: The Periodic Table at Work*. Vero Beach, FL: Rourke Publishing, 2008.

Cooper, Sharon Katz. *Periodic Table: Mapping the Elements*. Minneapolis, MN: Coughlan Publishing, 2007.

Jackson, Tom. *Radioactive Elements*. New York, NY: Marshall Cavendish, 2005.

Jerome, Kate Boehm. *Atomic Universe: The Quest to Discover Radioactivity*. Washington, DC: National Geographic Society, 2006.

Manning, Phillip. *Essential Chemistry: Atoms, Molecules, and Compounds*. New York, NY: Chelsea House Publishers, 2007.

McClafferty, Carla Killough. *Something Out of Nothing: Marie Curie and Radium*. New York, NY: Farrar, Straus and Giroux, 2006.

Miller, Connie. *Marie Curie and Radioactivity*. Minneapolis, MN: Coughlan Publishing, 2006.

Oxlade, Chris. *Atoms*. Chicago, IL: Heinemann Library, 2007.

Pettigrew, Mark. *Radiation*. North Mankato, MN: Stargazer Books, 2004.

Poynter, Margaret. *Marie Curie: Discoverer of Radium*. Berkeley Heights, NJ: Enslow Publishers, 2007.

Yannuzzi, Della. *New Elements: The Story of Marie Curie*. Greensboro, NC: Morgan Reynolds, 2006.

Bibliography

Argonne National Laboratory. "Natural Decay Series: Uranium, Radium, and Thorium." Retrieved April 19, 2008 (http://www.ead.anl.gov/pub/doc/natural-decay-series.pdf).

Baca, Bernadette. "How Dangerous Is Radium to Everyone?" Texas Department of Health–Bureau of Radiation Control. Retrieved April 19, 2008 (http://www.madsci.org/posts/archives/1999-03/920349021.Ch.r.html).

Carter, Laura Lee. "Glow in the Dark Tragedy." *American History Magazine*, October 2007. Retrieved April 18, 2008 (http://www.lauraleecarter.com/upload/AmHistRadiumCompleteCopySm.pdf).

Energy Information Administration. "U.S. Uranium Reserves Estimates." June 2004. Retrieved April 18, 2008 (http://www.eia.doe.gov/cneaf/nuclear/page/reserves/ures.html).

Fox, Margalit. "Ève Curie Labouisse, Biographer of Marie Curie, Dies at 102." *International Herald Tribune*, October 25, 2007. Retrieved April 18, 2008 (http://www.iht.com/articles/2007/10/25/europe/obits.php).

Frame, Paul. "William Crookes and the Turbulent Luminous Sea." Oak Ridge Associated Universities. Retrieved April 19, 2008 (http://www.orau.org/PTP/articlesstories/spinstory.htm).

Goldsmith, Barbara. *Obsessive Genius: The Inner World of Marie Curie.* New York, NY: Atlas Books, 2005.

Gray, Theodore. "For That Healthy Glow, Drink Radiation!" *Popular Science*, 265.2, August 1, 2004, p. 28. Retrieved April 19, 2008 (http://www.popsci.com/scitech/article/2004-08/healthy-glow-drink-radiation).

Health Canada. "How Would Radiation Affect My Body?" Retrieved
    April 19, 2008 (http://www.hc-sc.gc.ca/ed-ud/event-incident/
    radiolog/info/body-corps_e.html).

Kovarik, Bill. "The Radium Girls." Radford University. Retrieved April 18,
    2008 (http://www.runet.edu/~wkovarik/envhist/radium.html).

Nobel Foundation. "Wilhelm Conrad Röntgen." Retrieved April 18, 2008
    (http://nobelprize.org/nobel_prizes/physics/laureates/1901/
    rontgen-bio.html).

Occupational Safety & Health Administration. "Non-Ionizing Radiation."
    Retrieved April 19, 2008 (http://www.osha.gov/SLTC/radiation_
    nonionizing/index.html).

Parker, Barry. *Science 101: Physics*. New York, NY: HarperCollins
    Publishers, 2007.

PBS Online. "Chadwick Discovers the Neutron." Retrieved April 18, 2008
    (http://www.pbs.org/wgbh/aso/databank/entries/dp32ne.html).

Purdue University. "Ernest Rutherford." Retrieved April 18, 2008
    (http://chemed.chem.purdue.edu/genchem/history/rutherford.html).

Slowiczek, Fran, and Pamela Peters. "The Discovery of Radioactivity."
    Access Excellence at the National Health Museum. Retrieved April 18,
    2008 (http://www.accessexcellence.org/AE/AEC/CC/
    radioactivity.html).

Stwertka, Albert. *A Guide to the Elements*. 2nd ed. New York, NY:
    Oxford University Press, 2002.

University of Georgia Press. "Seven Natural Wonders of Georgia."
    Retrieved April 18, 2008 (http://www.georgiaencyclopedia.org/
    nge/Destination.jsp?id=p-54).

U.S. Environmental Protection Agency. "A Citizen's Guide to Radon."
    Retrieved April 19, 2008 (http://www.epa.gov/radon/pubs/
    citguide.html).

Waltar, Alan. *Radiation and Modern Life: Fulfilling Marie Curie's
    Dream*. Amherst, NY: Prometheus Books, 2004.

Index

About the Author

Kristi Lew is a professional K–12 educational writer with degrees in biochemistry and genetics. A former high school science teacher, Lew has written more than twenty books about science, health, and the environment for students and teachers.

Photo Credits

Cover, pp. 1, 16, 19, 22, 38–39 Tahara Anderson; p. 5 © NMPFT/DHA/SSPL/The Image Works; p. 7 The Art Archive/Culver Pictures; p. 9 Institut de Radium, Paris, France, Archives Charmet/The Bridgeman Art Library; p. 11 © Danita Delimont/Alamy; p. 13 University of Oxford, Museum of the History of Science, courtesy AIP Emilio Segrè Visual Archives; p. 15 © SSPL/The Image Works; pp. 20, 33 © Theodore Gray; p. 25 © C. Powell, P. Fowler & D. Perkins/Photo Researchers; p. 27 © Lawrence Berkeley National Laboratory; p. 29 © SSPL/The Image Works; p. 36 © Mary Evans Picture Library/The Images Works; p. 36 (inset) krtphotos/Newscom.com.

Designer: Tahara Anderson; **Editor:** Kathy Kuhtz Campbell
Photo Researcher: Amy Feinberg